Bibliographic information published by the German National Library:

The German National Library lists this publication in the National Bibliography; detailed bibliographic data are available on the Internet at http://dnb.dnb.de .

Imprint:

Print and binding: Books on Demand GmbH, Norderstedt Germany
ISBN: 9783668388093

This book at GRIN:

http://www.grin.com/en/e-book/342374/experimental-study-of-mrr-twr-sr-on-ss-316-and-aisi-d2-steel-using-aluminium

Sidhant Gupta

Experimental Study of MRR, TWR, SR on SS 316 and AISI D2 steel using Aluminium Electrode on EDM

GRIN Publishing

Experimental Study of MRR, TWR, SR on SS 316 and AISI D2 steel using Aluminium Electrode on EDM

A DISSERTATION SUBMITTED
IN PARTIAL FULFILLMENT OF THE REQUIREMENTS
FOR THE AWARD OF DEGREE OF

MASTER OF TECHNOLOGY
In
Mechanical Engineering
SUBMITTED BY

Sidhant

SUBMITTED TO

Ambala College of Engineering and Applied Research, Ambala

Kurukshetra University, Kurukshetra

2016

ACKNOWLEDGEMENTS

Dissertation work is an important aspect in the field of engineering. I express my sincere gratitude to Ambala College of Engineering and Applied Research, Ambala and Kurukshetra University, Kurukshetra for giving me the opportunity to work on the Dissertation during my final year of M.Tech.

I would like to thank my guide Dr. S.K. Jain, H.O.D, ME Department, and co-guide Er. Gurpinder Singh for their valuable support and to the members of Departmental Research Committee for their valuable suggestions and healthy criticism during dissertation work.
I would like to thank Dr. J.K. Sharma and Er. Manpreet Singh for their valuable support.
I would also like to thank everyone who has knowingly & unknowingly helped me throughout my Dissertation.
I am also thankful for the authors of all those books and papers which I had consulted during my Dissertation work as well as for preparing the report.

At the end, thanks to the Almighty for Everything.

Sidhant

ABSTRACT

In present study, the effect of aluminium tool electrode has been studied on stainless steel 316 and AISI D2 steel. Dielectric used for the study was EDM oil. Experiments were conducted based on L9 orthogonal array. The experimental study on the effect of input parameters i.e. current, pulse on time and pulse off time on output parameters material removal rate (MRR), tool wear rate (TWR) and surface roughness (SR). The workpiece materials selected were AISI D2 steel and SS 316. The tool electrode used was Aluminium and EDM oil as dielectric fluid. Taguchi design of experiments was used to design experiments, L9 orthogonal array was applied using MINITAB software. Signal to noise ratio and ANOVA were employed for parameter optimization and to achieve max MRR, min TWR and SR. The results indicate that the most influencing factor for MRR is Pulse off time. For TWR, the most influencing factor is current. For SR, the most influencing factor is pulse on time.

Contents

Chapter 2- Literature Review

Chapter 3- Methodology

Chapter 4 Analysis and Results

List of Figures

List of Tables

Abbreviations Used

EDM	Electric Discharge Machine
SS 316	Stainless Steel Grade 316
HCHCr	High Carbon High Chromium
MRR	Material Removal Rate
TWR	Tool Wear Rate
EWR	Electrode Wear Rate
SR	Surface Roughness
DC	Direct Current
HAZ	Heat Affected Zone
ANOVA	Analysis of Variance
S/N	Signal to Noise
Ton	Pulse on time
Toff	Pulse off time
µs	Micro Second
Seq. SS	Sequential Sum of Squares
Adj. MS	Adjusted Sum of Squares

1.1 Introduction and History of EDM:

When sparking takes place between two electrically conductive materials which are placed very close to each other, a small amount of material is removed from each of the material. It was first discovered by Joseph Priestly in 1770s. The Electric discharge machining started developing in mid 1970s. In mid 1980s, the EDM techniques were transferred to a machine tool. Today, it is the viable technique in the metal cutting industry with a wide number of applications and advantages.

The metal removal due to sparking was realized and attempts were made to harness and control the spark energy to employ it for useful purpose that is machining of metals. It was found that the spark of short duration and high frequency are needed for efficient machining. Further, it was also observed that if we submerge the discharge in dielectric, we can concentrate energy onto a small area.

A relaxation circuit that is RC circuit was proposed in which tool and workpiece i.e. electrodes are immersed into the dielectric like kerosene, and are connected to a capacitor. The capacitor is charged from a direct current source. Fig. 1.1 shows the RC circuit. As soon as the potential across the tool and the workpiece crosses the breakdown voltage, the sparking takes place at a point of least electrical resistance, which is usually the smallest inter-electrode gap (IEG). After every successive discharge capacitor recharges and spark will takes place at the next narrowest gap. Whenever a spark occurs heat is generated, which is shared in different modes by workpiece, tool, dielectric, debris and other parts of the system.

The dielectric serves some important functions like cools down the tool and workpiece, cleans the IEG, and concentrating the spark energy on a small cross sectional area.

EDM spark erosion is the same as having an electrical short that burns a small hole in a piece of metal it contacts. For a successful EDM process both the workpiece and tool must be electrically conductive.

The EDM process can be used in two different ways:

1. A pre-shaped electrode or tool usually made from copper or graphite or other conductive material is shaped in the same shape of the cavity required on the workpiece.

This tool is fed vertically down and the reverse shape of the tool is eroded into the solid workpiece.

2. A continuous travelling vertical wire electrode, the dia. of a small needle or less, is controlled by the computer to follow a path which is programmed to erode or to cut a slot or a narrow slot through the workpiece to produce the required shape

One of the most important factor in EDM operation is the removal of the metal particles that is chips from the working gap. This is known as flushing. Flushing these particles out of the gap between the workpiece to prevent them from forming bridges that cause short circuits.

Electric Discharge Machining is classified as Non-conventional machining process because the tool and the workpiece are not in direct contact, for the purpose of material removal.

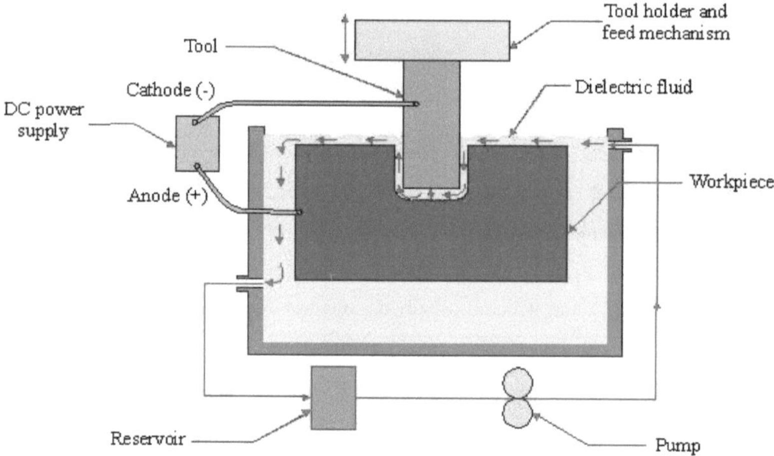

Fig 1.1 Schematics of EDM process

1.2 Working principle of EDM

Electric discharge machining (EDM) is a controlled metal removal process which is used to remove metals by electric spark erosion. In this process, electric spark is utilized as the cutting tool to cut the workpiece to produce the finished part of the desired shape. The metal removal process is carried out by applying a pulsating on/off electrical charge of high frequency current through the electrode of the workpiece. This removes very tiny pieces of metal from the workpiece at a controlled rate.

EDM is a thermo-electric process because heat energy of a spark is used to remove material from the workpiece. The workpiece and the tool should me electrically conductive materials. A spark is produced between the tool and the workpiece and the location of spark is determined by the narrowest gap between the two. Each spark is of very short duration. The time of the entire cycle is very few micro-seconds (μs). The frequency of the sparking may be as high as thousands of sparks per second. The spark is effective on a very small area. But the temperature of the spark area is very high. So the spark energy is capable of partly melting and partly vaporizing the material from the localized area of both tool and the workpiece.

The cavity produced on the workpiece is approximately the replica of the tool. But in order to achieve the exact replica of the tool, the tool wear must be zero. So to minimize the tool wear the operating parameters and polarity should be selected carefully. Particles eroded from the surface of electrodes are known as debris. Analysis of debris has revealed that it is a mixture of irregular shaped particles as well as hollow shaped particles.

A very small gap (of the order of hundredth of millimeter or even smaller) between the two electrodes is maintained in order to have a spark to occur. During EDM, pulsed DC of 80-100 V is passed through the electrodes. It results in the intense electrical field at the location where surface irregularity provides the smallest gap. Electrons break loose from the cathode surface and move towards the anode under the influence of electric field forces.

During this, the electrons collides with the neutral molecules of the dielectric and due to this electrons are also detached from these neutral molecules of the dielectric resulting in still more ionization. The ionization becomes so intense that a very narrow channel of continuous conductivity is established. In this continuous flow, considerable amount of electrons moves towards anode and that of ions moves towards cathode. Their kinetic energy is converted into heat energy hence heating of anode takes place due to bombardment of electrons and heating of cathode takes place due to bombardment of ions.

Thus it ends up in a momentary current impulse resulting in a discharge which is a spark. The spark energy raises the localized temperature of the electrodes to such a high value that it results in the melting, or melting as well as vaporization of material from the surface of both electrodes at the point of spark contact takes place in a very small amount.

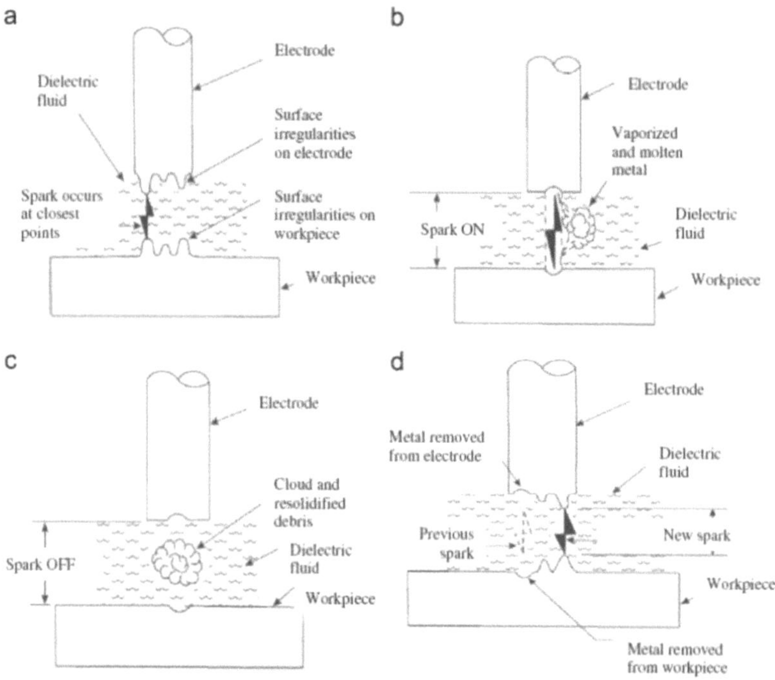

Figure No. 1.2(a) Occurrence of spark at the closest point between work piece and electrode, 1.2(b) melting and vaporization of work piece and electrode materials during spark on-time, 1.2(c) Vaporized cloud of materials suspended in dielectric fluid, and 1.2(d) Removal of molten metal and occurrence of next spark.

In fact, dielectric gets evaporated and the pressure in the plasma channel rises to a very high value and it does not let the evaporation of superheated metal. As soon as the off time of a pulse starts, the pressure drops instantaneously which allows the evaporation of superheated metal. The amount of material eroded from the workpiece and the tool will depend upon the contributions of K.E. of electrons and ions respectively. The polarity normally used is straight in which tool is negative and workpiece is positive while in reverse polarity vice versa. Movement of tool towards the workpiece is controlled by a servo mechanism. The sparking takes place over the entire surface of the workpiece hence the replica of the tool is produced. Usually a component made by EDM process is machined in two stages:

1. Rough machining at high MRR with poor surface finish,

2. Finish machining at low MRR with high surface finish.

1.3 EDM process

Among the thermal modes of machining, electrical discharge machining is mainly a technique used for the manufacture of a multitude of ever changing geometries very often produced as unit jobs or in small batches. After the pioneering investigations of Larezenko, the EDM process has attracted worldwide attention as a technique for metal machining and since then considerable research and development have been carried out.

The basic concepts of EDM process is crating out of metals affected by the sudden stoppage of the electron beam by the solid metal surface of anode. The portion of the anode facing the direct electrical pulse reaches the boiling point. Even in the case of medium long pulse the rate of temperature increase in tens of millions of degree per second which means dealing with an explosion process. The shock wave produced spreads from the center of the explosion to the inside the metal and on its way crushes the metal and deform crystal structure. In a very small duration of process the entire energy can only be expended in the surface layer of the anode. In reality, the mechanism of thermal conductivity has no time to start before the violet process of energy transfer is completed.

When a suitable unipolar (pulsed) voltage is applied across two electrodes by a dielectric fluid the latter breaks down. The electrons so liberated, are accelerated in presence of the electric field collide with the dielectric molecules, causing them to be robbed off their one or two electrons each and immunize.

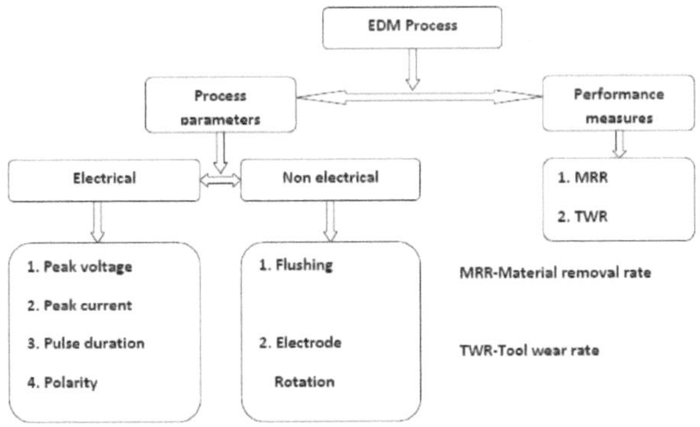

Fig 1.3 EDM process

The process grows and multiplies with secondary emission followed by an avalanche of electrons and ions. Dielectric layer's resistance drops as it is ionized resulting into ultimate breakdown. The electric energy is concentrated onto the gap and multifarious action takes place. Electro dynamic waves set in and travel at high speed high speed causing shock-impact and high temperature rise at the electrode surfaces. The instantaneous temperature may reach as high as 10,000 ^{0}C causing localized vaporization of the electrodes.

1.4 Advantages of EDM

To machine some of the hardest materials, EDM process has following important advantages which make it widely used in practice:

1. The process can be applied, in general, to any electrically conductive material. Other properties like strength, brittleness, etc. do not impose any restrictions to the application of the process.
2. The process provides a simple and straightforward method of form producing drop-forging, drawing and extruding dies and complex cavities in moulds and dies for plastics, die casting, glass and ceramic manufacturing.
3. Though the process involves temperature rise at the local spots to about 10,000 ^{0}C which can vaporize the localized material to machine, there is no heating of the bulk materials. However, the 'heat affected zone' (HAZ) surrounding the local points extends in the bulk to a depth of about few microns.
4. The high rates of heating and cooling at the treated surface renders some extra hardness (case-hardening) to the surface and this becomes a point of advantage in favour of the process.
5. Simple geometrical shape configurations can easily be produced by piercing or die-sinking in hardened die plates with required accuracy and surface finish. Thus, elimination of much complicated grinding and lapping is possible.
6. No mechanical stress is developed in the work material as there in no physical contact between the tool and work piece. This permits machining fragile and slender work pieces.
7. The process reduces time of machining in comparison with conventional grinding, honing or contour grinding etc.
8. The crater type non-directional (layless) surface pattern is said to retain lubricants, rendering the process particularly suitable for the finishing operation.

9. The surface finish produced by EDM process can be controlled to the required extent, minimizing the extra cost involved in additional operation for achieving improved surface finish.
10. Complex shapes which are difficult to produce with conventional machining methods can be produced on EDM
11. Extremely hard metals can be machined on EDM,
12. Very small work pieces which may get damaged from conventional cutting tool can be machined on EDM safely.
13. No direct contact between tool and the work pieces hence no mechanical stresses therefore delicate and weak materials can be machined without distortion.
14. Good surface finish can be achieved.

1.5 Disadvantages

1. The machining rate is very slow.
2. Potential fire hazard due to use of inflammable dielectrics use.
3. Very high power consumption.
4. Electrically non-conductive materials cannot be machined.
5. Overcut is formed.
6. Not capable to produce extreme sharp corners.
7. Size of the work piece is a constraint and depends on the machine set up. But generally large workpieces cannot be machined.

1.6 Dielectric Fluid

Since the process of removal of materials (both from work and tool) mainly depends upon thermal evaporation and melting, the presence of oxygen in atmosphere surrounding the spark would lead to formation of metal oxides which adversely affect the continuation or generation of repetitive sparks (most of metal oxides are bad conductor). Hence it is pertinent to use a dielectric fluid which contains no oxygen for liberation during the process, to help ionization, without disturbing the process.

But the performance of the dielectric to suit the purpose is extremely important. The failure of dielectric under electric stress, termed as breakdown, is found to spread over a wide range of applied stresses, depending upon its environment and mode of use.

1.7 Electrode material selection

1.7.1 Electrical Conductivity

The cutting tool in EDM is electric current, higher conductivity (or conversely, lower resistivity) promotes more efficient cutting.

1.7.2 Melting Point

We know that EDM is a thermal process, so we can assume that the higher the melting point of the electrode material, the better will be the wear ratio between electrode and workpiece.

1.7.3 Structural Integrity

EDM is thought of as a zero force process often, each individual spark is a very violent process on a microscopic scale, which exerts a considerable stress on the electrode material. How well the material surface will respond to these attacks, so it shall be an important factor in determining the electrode material's performance.

1.7.4 Mechanical Properties

The mechanical properties for electrode materials are:

1. Tensile strength
2. Transverse Rupture Strength (if applicable)
3. Grain Size (if applicable)
4. Hardness

1.7.5 Low Wear Burns

Electrode wear is a function of power supply settings, but is also of electrode properties. The combination of Electrode-workpiece pairs, polarity, Peak current and pulse on time, material lost from the electrode surface is re-plated back onto the electrode surface.

It should be noted that the re-deposited material is a combination of dielectric, workpiece and electrode.

Low wear rate is associated with electrode with positive polarity and long pulse on-time. Under extreme conditions, it is possible to actually grow the electrode rather than wear only. It is not a desirable condition.

1.7.6 Metallic Electrodes

The advantage of using metallic electrode is their electrical conductivity and mechanical integrity.

The disadvantages of using metallic electrodes are difficulty in fabrication and low cutting speeds.

1.8 EDM parameters

1.8.1 Pulse on time

Time during which machining takes place is called as pulse on time. Machining rate can be increased by increasing pulse on time and at the same time surface roughness also increases.

1.8.2 Pulse off time

Time during which reionization of the dielectric takes place is the pulse off time. Sufficient off time must be allowed before starting a new cycle. The shorter the pulse off time, faster will be the machining. But if pulse off time is kept too short then material removed from workpiece will not be swept away by the dielectric and dielectric will not deionize. This will cause the next spark to be unstable.

1.8.3 Peak Current

This is the most important parameter of the EDM, measured in amperes. As current increases material removal rate increases and current also affects electrode wear rate and surface finish. Higher the current, higher will be the tool wear rate and poor will be the surface finish. During the pulse-on time the current increases it reaches to level that is called peak current.

1.8.4 Voltage

It is potential that can be measured by volt. It affects material removal rate. Voltage is generally kept constant during experimentation. But it can be varied to study effects of voltage change on selected parameter.

2.1 Introduction

A lot of work has been done on different aspects of EDM. The EDM process has a wide application in die making industry and it also one of the best and most studied non-conventional machining technique. This chapter will cover the literature review of input parameters like Current, Pulse on time, Pulse off time, dielectric, and electrode on output parameters like Material removal rate, tool wear rate and surface roughness of the machined material. It is also a well suited method to machine hard to machined materials by conventional methods. Many parameters effect the quality of the machined material and hence many researchers had done work and gave their findings, a brief of which is given below.

2.2 Literature Survey

Lee and Li (2001) stated that for machining of tungsten carbide, graphite electrode gives the highest MRR with comparison to copper and copper tungsten electrode. For lower current values the wear ratio decreases for graphite but increases for copper. They exhibit copper tungsten exhibit lowest wear ratio at all ranges of peak current. They also observed the reverse polarity gives higher MRR, lower relative wear and better surface finish. They found optimal dielectric flushing pressure is 50Kpa. For precision machining of tungsten carbide, the optimum conditions of relative wear ratio and surface roughness achieved at a gap voltage of 120 V, discharge current 24A, pulse duration of 12.8 μs. [21]

Ho and Newman (2003) researched EDM relating to improving performance measures, optimizing the process variables. And monitor and control the sparking process. They suggested trend for future EDM research. [11]

Kristian (2004) states that EDM may be the only method available when manufacturing deep slots in low machinability materials. Their research concerns seal slots in Ni-Based turbine vanes. EDM Performance in respect to MRR and electrode wear is compared for 2 graphite qualities. They mentioned that based on the results the best choice of graphite quality grade for this app would be poco EDM3. If electrode wear also is important, very good conditions can be reached by some increase of pulse duration from 20 to 30 μs with a reduction of 6% MRR. [18]

Lauwers et al. (2004) have studied the material removal mechanisms for three composite ceramic materials, ZrO_2 based, Si_3N_4 based, Al_2O_3 with additions of electrical conductive phases like Tin and TiCN. The spalling effect is proven to be related with the formation of cracks. The formation of cracks depends on other factors like thermal conductivity of material, melting point & strength on the fracture toughness of material. [19]

Abbas et al. (2007) reviewed the research trends in EDM on ultrasonic EDM, dry EDM machining, EDM with powder additives, water EDM. They stated that ultrasonic vibration method is suitable for micro machining, dry machining is cost effective and EDM in water is introduced for safe and conductive working environment, EDM with powder additives is concerned more on improving surface quality. [1]

Khanra et al. (2007) stated that the low wear resistance of Cu, Cu alloys and graphite is a major problem. They developed a composite ZrB_2-Cu to get an optimum combination of wear resistance, electrical and thermal conductivity. This composite is used to machine mild steel. The ZrB_2 40% wt. Cu composite shows more MRR with less TWR than copper tool. But the diametral overcut and average surface roughness is found to be less in case of Copper tool than composite. [17]

Haron et al. (2008) observed the performance of copper and graphite tool electrode with XW42 tool steel. They observed material removal rate by copper and graphite electrodes of 10, 15, 20 mm dia. By varying current and machining time one by one. MRR of XW42 tool steel with copper is greater than with graphite electrode. Copper is suitable for roughing process, while graphite electrode is suitable for finishing. They concluded that MRR not only depends upon the dia. Of electrode and supply of current but also on the type of electrode material used. [9]

Amin et al. (2009) discussed the feasibility of machining Tungsten carbide ceramics with a graphite electrode on EDM using Taguchi method. Taguchi method has been used to determine the optimum machining conditions for the performance of EDM. They concluded that, the peak current affects the electrode wear rate and surface roughness largely. The pulse duration mainly affects the MRR. By using ANOVA table they have found that peak current is most significant factor for EWR and SR. [2]

Sameh (2009) developed a comprehensive mathematical model for correlating the various EDM parameters and their influence through response surface methodology. They selected conductive metal matrix composite Al/Si as workpiece. Copper is used as electrode. They concluded that with increases in pulse on time cause as increase in MRR until it reaches 200 μs

& then it began to decrease. EWR decreases as pulse on time increases for a combination of gap voltage and peak current. As MRR decreases, EWR also decreases. The surface roughness of the machined surface increases as the energy as the pulse energy increases. It means, MRR increases at a high pulse energy and hence surface will be rough. [31]

Jegan et al. (2012) investigated the EDM parameters using gray rational analysis on AISI 202 stainless steel, parameters mainly discharge current, and pulse on time and pulse off time and optimization of these parameters are done by GRA. The purpose of this experimental work is to know the effect of machining parameters and conclude these effects on MRR and SR. EDM oil is used as dielectric and workpiece AISI202 and copper as electrode. Their results shows that the main parameter that effect MRR is discharge current, so by properly adjusting the control factors, quality of product can be improved. [14]

Gopalakannan et al. (2012) carried out an experimental investigation to study the effect of pulsed current on MRR, EWR and SR and diametral overcut in SS 316L and 17-4 PH, electrodes used were copper, copper tungsten, graphite. It is observed that output parameters such as MRR, EWR and SR increases with increase in pulsed current. High MRR achieved with copper electrode whereas copper tungsten yielding lower EWR, smooth SF and good dimensional accuracy. [7]

Hindus et al. (2013) investigated the machining characteristics of SS 316L through EDM. Copper is used as electrode. Results indicate that MRR and TWR is strongly influenced by current and pulse on time. Most significant factor for MRR found to be pulse on time followed by current. For TWR the most significant factor was current followed by pulse on time. [10]

Murikan et al. (2013) investigated environmental friendly dry EDM. Liquid dielectric is replaced by gaseous dielectrics. Investigation is done on stainless steel 316L using compressed air as dielectric and copper electrode as tool. The influence of discharge current, pulse on time, duty factor and spindle speed on MRR and TWR has been studied. From the results it is observed that maximum MRR of 9.94 mm^3/min is obtained and is influenced by discharge current, pulse on time and duty factor. Minimum tool wear of 0.048 mm^3/min is obtained and tool wear is influenced by discharge current and duty factor. [24]

Patel et al. (2013) performed experiments to determine parameters effecting surface roughness. Experiments were using copper, brass and aluminium as tool electrodes on Mild steel and kerosene oils as dielectric. MRR obtained: copper; 1.45 gm (10μs, 23.48 A), brass; 0.7 gm (200 μs, 23.48 A) and aluminium; 1.48 gm (20 μs, 23.48 A). SR increased with increase in current. SR of aluminium is more than copper followed by brass. [27]

Choudhary et al. (2013) investigated the effect of different electrodes on MRR, SR of SS 316. Electrode material, current and pulse on time were taken as variables for the study of MRR and SR. Copper, brass and graphite were used as electrodes and EDM oil as dielectric. They concluded for MRR, electrode material is the most influencing factor and then discharge current. MRR increases with increase in current. Copper electrode shows highest MRR while Brass showed less. SR is better with lower value of current. Brass shows the better surface finish while copper shows the worst surface finish. [3]

Rahi et al. (2014) parametric analysis is conducted on High carbon high chromium steel with copper and graphite as electrodes and high carbon oil as dielectric. The effect of input parameters on MRR, EWR, SR has been studied. For MRR with copper electrode input current and duty cycle is more dominant. Pulse on time and duty cycle for EWR, duty cycle and pressure for surface roughness. [28]

Rajendra M. et al. (2014) investigated the effect of abrasive mixed dielectric on High carbon high chromium steel D3. Abrasive material used was Al_2O_3. Results indicates that abrasive particle size and concentration and pulse current are the most significant parameters that improves MRR. By adding abrasive particle, efficiency of EDM increases. As per S/N ratio and ANOVA, MRR is influenced by discharge current and abrasive concentration. At 6 g/ltr of concentration MRR is maximum. Abrasive mixed EDM results in 58% more MRR than the traditional EDM. [29]

Laxman and Guru Raj (2014) optimized the EDM parameters using gray relational analysis. Effect of parameters peak current, pulse on time, and pulse off time on MRR and TWR is studied. Taguchi design of experiments is used. Dielectric used was EDM oil grade 30. Copper electrode is used for experimentation. Experimental investigation has been carried out on titanium super alloys. The results of ANOVA indicates that pulse current is the most influencing factor for machining of titanium super alloy. [20]

Raju et al. (2014) investigated the effect of surface roughness of the most influenced parameters pulse on time, peak current, servo voltage and wire tension on SS 316L. Signal to noise ratio is used to find the optimal combination of parameters. Assumptions of ANOVA were verified and found to be valid. It is found that the machining variable pulse on time is the significant factor that effect surface roughness. [30]

Suresh et al. (2014) conducted investigation of MRR of SS316L. Input parameters selected are current, pulse on time and pulse off time. Tungsten electrode was used of 300 μm diameter. The S/N ratio on MRR reveals that the current plays the most significant role in the parameters chosen thus the developed mathematical model is well suited to improve the productivity for selecting the optimal parameters. [36]

Dixit et al. (2015) investigated the parameters MRR, EWR on AISI D3 steel. Input parameters are pulse on time, pulse off time, peak current and fluid pressure. Copper electrode is used to perform the experiment and analyzed using Taguchi method. It is concluded that MRR is mainly influenced by peak current and EWR is mainly influenced by peak current followed by pulse on time. [6]

2.3 Gap in Literature

From literature survey it is seen that a lot of work has been done on different aspects of EDM. Many researchers worked on EDM parameters like current, pulse on time, pulse off time, electrode material and dielectric fluid and studied the effect of these parameters on material removal rate, tool wear rate and surface roughness. Many researchers have used different electrodes like copper, brass and graphite. Few have used tungsten and tungsten carbide. But copper and graphite being the most commonly used due to the fact that conductive properties are good. Very few researchers have used aluminium as electrode. Aluminium also has very good conductive properties. As we can see in literature review, materials used for research purpose are SS 316, mild steel, AISI 202, titanium alloys and AISI D2 and D3 steel. SS 316 has wide number of applications because of its very high corrosion resistance. Now a days it is used in orthopedic implants and also in Food preparation equipment particularly in chloride environments, pharmaceuticals piping. SS 316 is hard to machine material with conventional machining processes. So EDM is a suitable method.

From the literature review it is concluded that no work has been reported on aluminium as electrode on SS 316 and AISI D2 steel. Copper, brass and graphite are mostly used. Aluminium is also a very good conductor of electricity and also cheaper than copper, brass and graphite. So this might help in improving productivity. For this study. Workpiece taken are Stainless steel grade 316 steel and high carbon high chromium steel. Dielectric used is EDM oil. The experimental design has been performed on MINITAB 17 software. The results has been analyzed by ANOVA.

2.4 Objective of the present study

1. To determine the effect of input parameters current, pulse on time and pulse off time on output parameters MRR, TWR and SR.
2. To investigate the effect on MRR, TWR, SR by using aluminium electrode on SS 316 and AISI D2 steel.
3. Prediction of optimal values of input parameters using MINITAB.

3.1 Phasing of the work

This chapter explains the methodology followed for the research work. Experiments were conducted based on Taguchi's method and L9 orthogonal array has been used for machining of both SS 316 and High carbon high chromium AISI D2 steel. Three input parameters namely current, pulse on time and pulse off time having three levels each. MRR, TWR and SR has been measured and studied. The MINITAB 17 SOFTWARE is used for analyzing the Taguchi Design of experiments. The procedure adopted for experimental study is represented as follows:

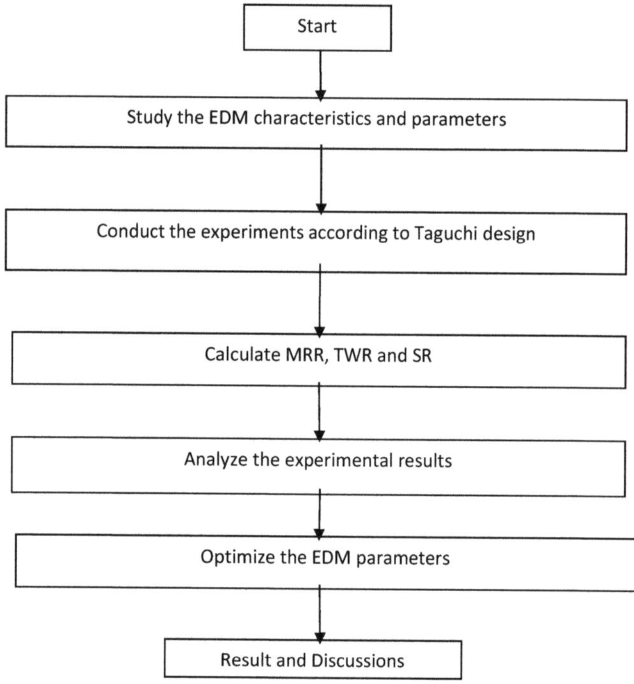

Figure 3.1: Procedure for the research work

3.2 Experimental set up

The experimentation is done according to Taguchi method as per L9 orthogonal array with three controllable parameters. Experiments on EDM was conducted in CITCO, Chandigarh. The materials to be machined was SS316 and AISI D2 steel using aluminium electrode of diameter 15 mm and EDM oil as dielectric. In the present work response variables Material removal Rate (MRR), Tool Wear Rate (TWR) and Surface Roughness (SR) has been measured and studied. Material is removed every time from the workpiece due to heat generated by the arc, material is also wear out from the electrode as well. So material removal is measured after every run of experiment for both electrode and workpiece. Material removal from electrode is termed as tool wear rate or electrode wear rate. For each set of experiments, observation tables were made and according to these observation tables results were being calculated.

3.3 Workpiece and Electrode used in the experimentation work

3.3.1 Workpiece

Stainless steel grade 316 and high carbon high chromium steel are very hard materials and also have very high corrosion resistance. Due to this they have a wide number of applications. Applications like food preparation equipment, pharmaceutical industry. Medical implants, fasteners, and marine industry. The SS316 workpiece was of 50 mm dia. and 5 mm in thickness.

Table 3.1 – SS 316 composition

Composition of SS 316								
C	Mn	Si	Cr	Ni	Mo	P	S	Fe
0.08%	2.00%	0.75 to 1.00 %	18.00%	14.00%	2.00 to 3.00%	0.05%	0.03%	Balance

Figure 3.2 - Workpiece before Machining (SS316)

Figure 3.3 - Workpiece after Machining (SS316)

High carbon high chromium steel is used mainly in die and mould making industry. Now a days dies are of very complicated shape so to achieve that kind of shape, non-conventional machining methods are employed because it cannot be achieved by conventional methods.

Table 3.2 – AISI D2 steel Composition

Composition of AISI D2 steel								
C	Mn	Si	Cr	Ni	Mo	P	S	Fe
1.59%	0.20%	0.40%	11.60%	0.24%	0.11%	0.05%	0.09%	Balance

Figure 3.4 - Workpiece before machining (AISI D2 steel)

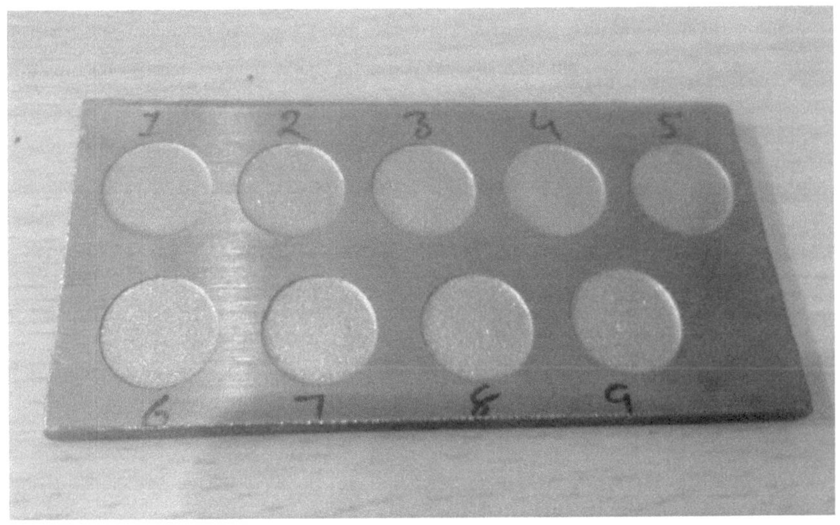

Figure 3.5 - Workpiece after Machining (AISI D2 steel)

Electrode used for the current research work is Aluminium. Aluminium as the conductivity 3.50×10^{7} S/m at 20^{0} Celsius. And temperature co-efficient (K^{-1}) 0.0039 almost comparable to copper. Depending on these properties Aluminium can be considered as a suitable EDM electrode and also there is no research work stated on Aluminium electrode.

3.3.2 Electrode Material for experimentation work

Table 3.3 – Aluminium Properties

Aluminium Properties	
Melting Point	933.47 K
Boiling Point	2743 K
Density	2.70 g/Cm3
Thermal Expansion at25 degree C	23.1 µm/m.K
Thermal Conductivity	237 W/m.K
Brinnell Hardness	160-550 MPa
Young's Modulus	70 GPa
Shear Modulus	26 GPa
Crystal structure	FCC
Electrical Resistivity	2.82×10^{-8}
Electrical Conductivity	3.50×10^{7}

Aluminium is silvery white, soft, non-magnetic, ductile metal. It is third most abundant element in the earth's crust, the most abundant metal. The chief ore is bauxite. Remarkable for the metal's low density and its ability to resist corrosion. Aluminium and its alloys are vital to aerospace industry. Aluminium is capable of super conductivity. Aluminium used in electrical transmission lines for power transmission. Diameter of the electrode was 15mm. Table 3.3 shows the properties of Aluminium metal.

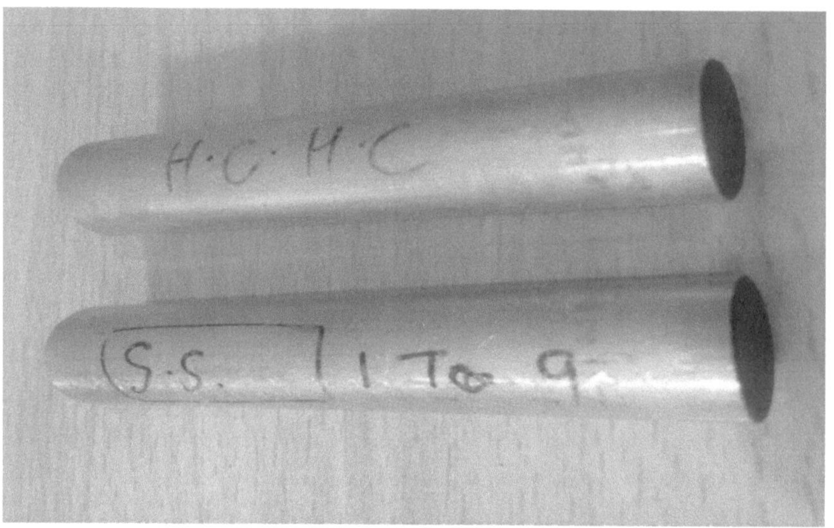

Figure 3.6 Aluminium Electrodes used for experiments

3.4 Machine used for experiments

The machine was used die sinker EDM brand Elektra model number EMS 5535. It consists of following major parts:

- Dielectric reservoir and pump circulation system
- Power generator and control unit
- Working tank with work holding device
- x-y table accommodating the working table
- tool holder

3.5 Specifications of EDM

Table 3.4 – EDM specifications

Working current	50amp
Open gap voltage	100 V
Pulse on time	1 to 2000 μs
Pulse frequency	0.22 to 330 kHz
Maximum stroke removal copper and steel	300mm^3/min
Input main voltage	415 3 phase , 50hz balanced
Power factor	Approx 0.9 lag
Power consumption	Approx 7.5 kVa
Weight	300 kg
Height	1320mm
Width	725mm
Depth	450mm

3.6 Machine used to measure weight

For the measurement of MRR and TWR and for calculations, we need to weight workpiece and electrode both before and after each experiment. To measure the initial and final weight of the workpiece after every run. The machine used was off brand Wenscer and model PSG 300. The resolution of the machine was 0.01gm.

3.7 Machine used for surface roughness

Fig 3.7 Surface roughness tester

The surface roughness tester machine used was SRT 6210.When measuring the roughness of a surface, the sensor is placed on the surface and then uniformly slides along the surface by driving the mechanism by the sharp built-in probe. This roughness causes displacement of the probe which results in change of inductive amount of induction coils so as to generate analogue signal, which is in proportion to the surface roughness at output end of phase-sensitive rectifier. The exclusive DSP processes and calculates and then outputs the measurement results on LCD. Specifications are given in Table 3.5.

Table 3.5 – SRT 6210 specifications

Test Principle	Inductance type
Radius of Probe Pin	5
Material of Probe Pin	Diamond
Dynamo-measurement of Probe	4mN(0.4gf)
Probe Angle	90
Vertical Radius of Guiding Head	48mm
Weight	about 420 g
Power Li-ion battery	rechargeable
Operating conditions	Temp. 0~50°C, Humidity <80%

3.8 Taguchi Method

In the current investigation, a total of 18 experiments were conducted based on Taguchi's method. As per L9 orthogonal array, three levels are considered for each factor. As shown in table 3.6.

The S/N ratio and ANOVA was determined for finding the most significant factors affecting the machining performance i.e. material removal rate, electrode wear rate and surface roughness. In the software we have added all the experimental values and got the optimized result. In MINITAB, Taguchi design recognizes that not all factors that cause variation can be controlled. Those uncontrolled factors are called noise factors. It recognizes controllable factors. Taguchi Design use orthogonal array, which estimate the effect of factors on response mean and variation. During experimentation, you manipulate noise factor to force variability to occur and then determine optimal control factors. An orthogonal array means the design is balanced so that factor levels are weighted equally. Because of this factors can be assessed independently of all other factors. MINITAB calculates response tables, main effects and interaction plots:

4. S/N Ratio vs the controllable factors,
5. Standard Deviation vs the controllable factors,
6. Natural Log of standard deviation vs control factors,

Delta is the difference between maximum and minimum average standard deviation for the factors. The rank is the rank of each delta, where rank is largest delta. Mean is the average response for each combination of control factors level in static Taguchi Design. Depending on response, your goal is to determine factor levels that either minimize or maximize the mean.

3.8.1 Input Parameters

Following table shows the parameters selected for the present work and their respective levels. Three values are selected for each parameters for all three levels. While selecting the parameters there were some constraints of the machine. So the values what were thought earlier was changed.

Table 3.6 – Input Parameters and the their levels

S. No.	Input Parameters	Level		
		1	2	3
1	Current (A)	6	8	10
2	Pulse On (Ton)	50	100	150
3	Pulse Off (Toff)	6	8	10

3.8.2 Response variables

In the present study, Material removal rate, electrode wear rate and surface roughness are the three main variables which are measured and studied.

3.9 Evaluation of MRR

MRR is expressed as the ratio of the difference of weight of the work piece before and after machining to the machining time. It is measured in mg/min.

$$MRR = (M_i - M_f)/t$$

Whereas, M_i = Weight of work piece before machining i.e. initial weight

M_f = Weight of work piece after machining i.e. final weight

3.10 Evaluation of EWR

EWR is expressed as the ratio of the difference of weight of the electrode before and after Machining to the machining time.

$$EWR = E_i - E_f / t$$

Whereas, E_i = Weight of the electrode before machining.

E_f = Weight of the electrode after machining.

4.1 Analysis and result of MRR of AISI D2 steel

Total 9 experiments were conducted for the L_9 orthogonal array for both the materials. The results of each experiment were repeated three times, which means for every experiment there are three values of material removal rate. So the analysis is based on it. But to calculate results of MRR, S/N ratio is to be calculated first.

4.1.1 Analysis and result of MRR

The effect of input parameters i.e. current, Pulse on time, pulse off time was evaluated using ANOVA. Signal to noise ratio is calculated and the response data specially used for design of experiments application is analyzed.

(S/N) HB = -10 log (MSDHB)

Where MSDHB=1/r$\sum_i^r = (\frac{1}{y_i^2})$

r = the number of tests in a trial

y_i = observed value of response characteristics

For material removal rate (MRR) the S/N ratio is larger is better.

Where, MSDHB = Mean Square deviation for higher the better response.

4.1.2 Observation Table for MRR

While conducting all the 9 experiments with different values of input parameters, observations were made, Table no. 4.1. Weight lost per gram from the workpiece in each experiment and time taken for each experiment was also observed. The depth of cut was fixed to 0.5 mm and time taken to achieve this cut was observed. For some experiments time was higher and for some experiments time is low. So we calculated an average material removal rate for 10 minutes. And after each experiment the weight of workpiece is measured and loss in weight is observed as removal of material. After completion of all experiments the observation table is made by filling all the values.

$$\text{MRR} = \frac{\text{Final weight} - \text{Initial weight}}{\text{Time of machining}}$$

Table 4.1 Observation Table for Material Removal Rate of AISI D2 steel

Serial No	Current Ip, A	Pulse on Ton, μs	Pulse off Toff, μs	Weight of workpiece (Grams)			SNRA1	MEAN1
				MRR1	MRR2	MRR3		
1	6	50	6	0.1195	0.1232	0.1211	-18.32719978	0.121266667
2	6	100	8	0.3411	0.3144	0.3356	-9.636345523	0.330366667
3	6	150	10	0.72	0.7329	0.7469	-2.697682871	0.733266667
4	8	50	8	0.36	0.3713	0.3675	-8.726206938	0.366266667
5	8	100	10	1.0166	1.0087	1.021	0.13270037	1.015433333
6	8	150	6	0.2031	0.1967	0.2104	-13.84282491	0.2034
7	10	50	10	0.7	0.6934	0.7062	-3.100420712	0.699866667
8	10	100	6	0.5166	0.5251	0.5286	-5.623987052	0.523433333
9	10	150	8	0.57	0.5772	0.5658	-4.868160416	0.571

4.1.3 Analysis of Variance for MRR

The results were analyzed using ANOVA for identifying the significant factors affecting performance. The variation data for each factor and their interactions were tested to find significance of each calculated by formula. Main effect plot is shown in fig. 4.1, shows the variation of MRR with input parameters.

The ANOVA table for Mean MRR at 95% confidence level. The data for each factor and its interaction were F tested, Where F is the Fister Value. The greater its effect on the performance characteristics. The corresponding values in ANOVA table are shown in table 4.4. This table shows that pulse off time, current, pulse on time affects the MRR significantly. Pulse off has the highest contribution to MRR.

4.1.4 Confirmation test for MRR

From mean of each level of every factor, response table is constructed for MRR table no 4.2 and 4.3. From Table it is concluded that for MRR, the optimum conditions are A3, B2 and C3 i.e. current 10 A, Pulse on 100 μs and pulse off 10 μs. From figure 4.1 and 4.2, shows the main effect plot for S/N ratio and Means and obtaining optimal values.

Graph show that larger the value of parameters better the MRR. Here A is current; max at A3, pulse on; max at B2, pulse off C3. The ANOVA table is shown in table 4.4.

Table 4.2 – Response table for S/N ratio of D2 Steel

Response table for S/N ratio (Larger is Better)			
Level	Current (A)	Pulse On (B)	Pulse off (C)
1	-10.22	-10.051	-12.598
2	-7.479	-5.043	-7.744
3	-4.531	-7.136	-1.888
Delta	5.69	5.009	10710
Rank	2	3	1

From graph 4.1 and 4.2, main effects plot for MRR conclude that optimum condition for MRR is Current (10 A), Pulse on (100 μs), Pulse off (10 μs). The main effects plot shows the values of parameters and also showing optimal values of parameters. Plots are explained in the respective heading. From plots we can conclude the optimum value of the parameters to make the process robust.

Table 4.3 – Response Table for Means of D2 Steel

Response Table for Means			
Level	Current (A)	Pulse on (B)	Pulse off (C)
1	0.39	0.3958	0.2827
2	0.5284	0.6231	0.4225
3	0.5981	0.5026	0.8162
Delta	0.2031	0.2273	0.5335
Rank	3	2	1

Table 4.4 – ANOVA table for S/N ratio

Analysis of Variance for SN ratio					
Source	DF	Seq. SS	Adj. MS	F	P
Current	2	48.58	24.289	3.14	0.242
Pulse on	2	37.97	18.984	2.45	0.29
Pulse off	2	172.54	86.271	11.14	0.082
Residual Error	2	15.49	7.744		
Total	8	274.58			

Table 4.5 – ANOVA table for Means

Analysis of Variance for Means					
Source	DF	Seq. SS	Adj. MS	F	P
Current	2	0.06392	0.03196	1.5	0.4
Pulse on	2	0.07758	0.03879	1.82	0.355
Pulse off	2	0.45912	0.22956	10.77	0.085
Residual Error	2	0.04261	0.02131		
Total	8	0.64323			

4.1.5 Main effects plots for MRR of AISI D2 steel

We can see that the graphs are showing minimum as well as maximum values of parameters. For MRR, we chosen ANOVA as larger is better so here we are interested in maximum values. There are two plots; one is for SN ratio and other is for means. For SN ratio we are always interested in maximum values because it tells us which parameter at which value has the maximum effects on the process. Whereas for means plot it depends on ANOVA, here larger is better so larger are considered to optimize the process.

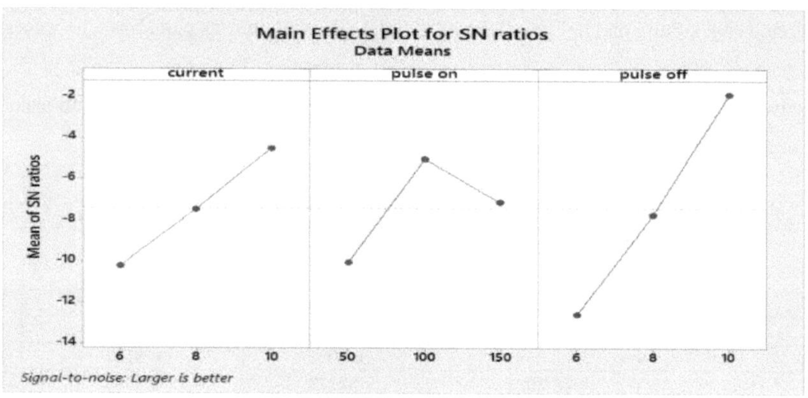

Figure 4.1–Main effects Plot for SN ratio,

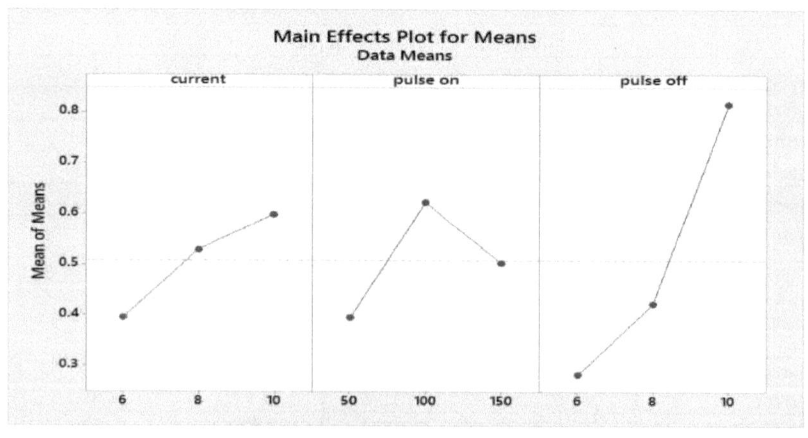

Figure 4.2 - Main Effects Plot for Means

4.2 Results and Analysis of TWR of AISI D2 steel

Total 9 experiments were conducted for the L_9 orthogonal array for both the materials. The results of each experiment were repeated three times, which means for every experiment there are three values of material removal rate. So the analysis is based on it. But to calculate results of TWR, S/N ratio is to be calculated first.

4.2.1 Results and analysis of TWR

The effect of input parameters i.e. current, Pulse on time, pulse off time was evaluated using ANOVA. Signal to noise ratio is calculated and the response data specially used for design of experiments application is analyzed.

(S/N) LB = -10 log (MSDLB)

Where, MSDLB=$1/r\sum_{i=1}^{r} = (y_i^2)$,

r = the number of tests in a trial

y_i = observed value of response characteristics

For Tool wear rate (TWR) the S/N ratio is smaller is better.

Where MSDLB = Mean Square deviation for smaller the better response.

4.2.2 Observation Table for TWR

Table 4.6 – Observation Table for Tool Wear Rate of AISI D2 Steel

Serial No	Current Ip, A	Pulse on Ton, µs	Pulse off Toff, µs	Weight of electrode (Grams)			SNRA1	MEAN1
				TWR1	TWR2	TWR3		
1	6	50	6	0.0268	0.0234	0.0274	31.72510106	0.025866667
2	6	100	8	0.0411	0.0489	0.0477	26.73957182	0.0459
3	6	150	10	0.03	0.0392	0.0342	29.20061811	0.034466667
4	8	50	8	0.0769	0.0648	0.0698	23.01473492	0.0705
5	8	100	10	0.06	0.0567	0.0622	24.48398351	0.059633333
6	8	150	6	0.0269	0.0212	0.0287	31.76799737	0.0256
7	10	50	10	0.133	0.1411	0.1369	17.263058	0.137
8	10	100	6	0.05	0.0478	0.0493	26.18865056	0.049033333
9	10	150	8	0.05	0.0478	0.0493	26.18865056	0.049033333

While conducting all the 9 experiments with different values of input parameters, observations were made. Weight lost per gram from the electrode in each experiment and time taken for each experiment was also observed. The depth of cut on workpiece was fixed to 0.5 mm and time taken to achieve this cut was observed. For some experiments time was higher and for some experiments time is low. So we calculated an average tool wear rate for 10 minutes. And after each experiment the weight of tool electrode is measured and loss in weight is observed as removal of material. After completion of all experiments the observation table 4.6 is made by filling all the values.

4.2.3 Analysis of Variance

The results were analyzed using ANOVA for identifying the significant factors affecting performance. The variation data for each factor and their interactions were F tested to find significance of each calculated by formula. Main effect plot is shown in fig. 4.3, shows the variation of TWR with input parameters. The ANOVA table for Mean TWR at 95% confidence level. The data for each factor and its interaction were F tested, Where F is the Fister Value. The greater its effect on the performance characteristics. The corresponding values in ANOVA table are shown in table 4.9. This table shows that pulse off time, current, pulse on time affects the TWR significantly. Pulse off has the highest contribution to TWR.

4.2.4 Confirmation test for TWR

From mean of each level of every factor, response table is constructed for TWR table no 4.7 and 4.8. From Table it is concluded that for TWR, the optimum conditions are A1, B3 and C1 i.e. current 6 A, Pulse on 150 μs and pulse off 6 μs. From figure 4.3 and 4.4, shows the main effect plot for S/N ratio and Means and obtaining optimal values.

Graph show that larger the value of parameters better the TWR. Here A is current; max at A1, pulse on; max at B3, pulse off C1. The ANOVA table is shown in table 4.9

Table 4.7 Response table for S/N ratio

Response table for S/N ratio			
Level	Current (A)	Pulse On (B)	Pulse off (C)
1	29.22	24	29.89
2	26.42	25.8	25.31
3	23.21	29.05	23.65
Delta	6.01	5.05	6.24
Rank	2	3	1

From graph 4.3 and 4.4, main effects plot for TWR conclude that optimum condition for TWR is Current (6 A), Pulse on (150 μs), Pulse off (6 μs). The main effects plot shows the values of parameters and also showing optimal values of parameters. Plots are explained in the respective heading. From plots we can conclude the optimum value of the parameters to make the process robust.

Table 4.8 Response Table for Means

Response Table for Means			
Level	Current (A)	Pulse On (B)	Pulse off (C)
1	0.03541	0.07779	0.0335
2	0.05191	0.05152	0.05514
3	0.07836	0.03637	0.07703
Delta	0.04294	0.04142	0.04353
Rank	2	3	1

Table 4.9 – ANOVA table for S/N Ratio

Analysis of Variance for SN ratio					
Source	DF	Seq. SS	Adj. MS	F	P
Current	2	54.233	27.117	7.31	0.12
Pulse on	2	39.32	19.66	5.3	0.159
Pulse off	2	62.742	31.371	8.45	0.106
Residual Error	2	7.432	3.711		
Total	8	163.718			

Table 4.10 – ANOVA table for Means

Analysis of Variance for Means					
Source	DF	Seq. SS	Adj. MS	F	P
Current	2	0.002816	0.001408	2.87	0.258
Pulse on	2	0.002635	0.001318	2.69	0.271
Pulse off	2	0.002843	0.001421	2.9	0.256
Residual Error	2	0.00098	0.00049		
Total	8	0.009274			

For TWR, ANOVA is smaller the better. From table 4.7 and 4.8, it is concluded that pulse off is the most significant factor for TWR, followed by current and then pulse on. So pulse on is the least significant factor.

From graph 4.3 and 4.4, main effects plot for TWR conclude that optimum condition for TWR is Current (6 A), Pulse on (150 μs), Pulse off (6 μs). The main effects plot shows the values of parameters and also showing optimal values of parameters.

4.2.5 Main effects plots

We can see that the graphs are showing minimum as well as maximum values of parameters. For TWR, we chosen ANOVA as smaller is better so here we are interested in minimum values. There are two plots; one is for SN ratio and other is for means.

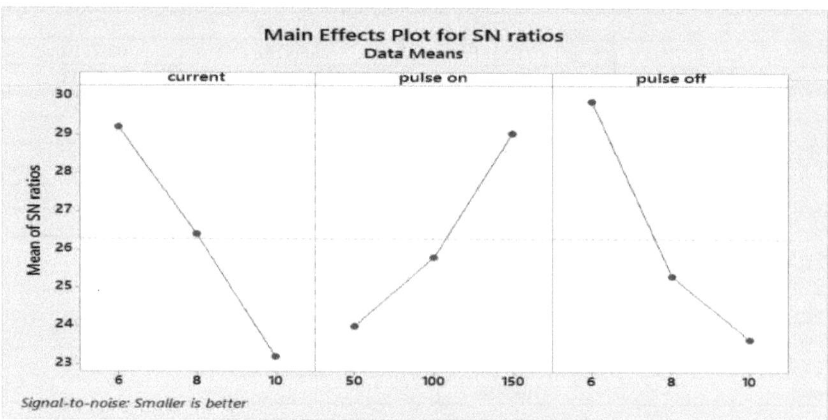

Figure 4.3-Main effects plot for SN ratio,

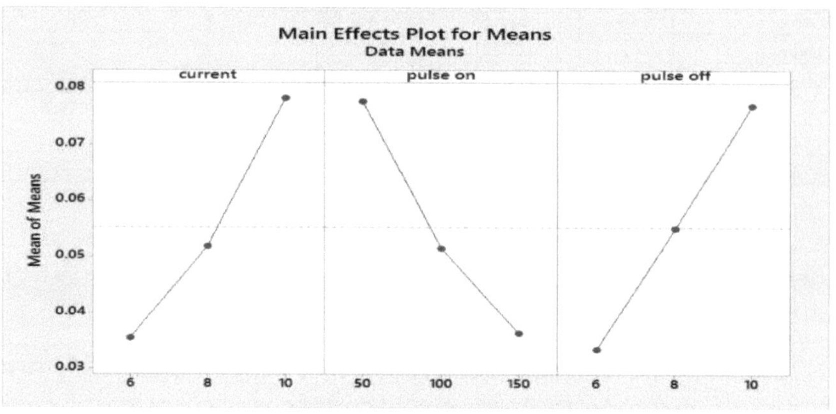

Figure 4.4-Main Effects plot for Means

4.3 Results and Analysis of Surface Roughness of AISI D2 Steel

Total 9 experiments were conducted for the L_9 orthogonal array for both the materials. The results of each experiment were repeated three times, which means for every experiment there are three values of surface roughness. So the analysis is based on this experimentation table. To calculate results of SR, S/N ratio is to be calculated first.

4.3.1 Results and analysis of Surface roughness

The effect of input parameters i.e. current, Pulse on time, pulse off time was evaluated using ANOVA. Signal to noise ratio is calculated and the response data specially used for design of experiments application is analyzed.

(S/N) LB = -10 log (MSDLB)

Where, MSDLB=1/r$\sum_{i=1}^{r} = (y_i^2)$,

r = the number of tests in a trial

y_i = observed value of response characteristics

For Tool wear rate (TWR) the S/N ratio is smaller is better.

Where MSDLB = Mean Square deviation for smaller the better response.

4.3.2 Observation Table for SR

While conducting all the 9 experiments with different values of input parameters, observations were made.

Table 4.11 - Observation Table for SR of AISI D2 Steel

Serial No	Current Ip, A	Pulse on Ton, µs	Pulse off Toff, µs	Surface Roughness				
				SR1	SR2	SR3	SNRA1	MEAN1
1	6	50	6	3.518	3.62	3.578	-11.05882531	3.572
2	6	100	8	5.472	5.523	5.344	-14.72291437	5.446333333
3	6	150	10	5.784	5.821	5.987	-15.36486332	5.864
4	8	50	8	3.961	3.824	3.745	-11.69650162	3.843333333
5	8	100	10	5.238	5.298	5.327	-14.46549474	5.287666667
6	8	150	6	6.618	6.719	6.439	-16.38168346	6.592
7	10	50	10	4.977	5.003	4.829	-13.86913615	4.936333333
8	10	100	6	6.618	6.724	6.784	-16.53317877	6.708666667
9	10	150	8	6.644	6.731	6.678	-16.50128655	6.684333333

Weight lost per gram from the workpiece in each experiment and time taken for each experiment was also observed. The depth of cut was fixed to 0.5 mm and time taken to achieve this cut was observed. For some experiments time was higher and for some experiments time is low. So we calculated an average material removal rate for 10 minutes. And after completion of experimental work, Surface roughness is measured of these cuts. After completion of all experiments the observation table 4.11 is made by filling all the values.

4.3.3 Analysis of Variance of SR

The results were analyzed using ANOVA for identifying the significant factors affecting performance. The variation data for each factor and their interactions were tested to find significance of each calculated by formula. Main effect plot is shown in fig. 4.5, shows the variation of SR with input parameters. The ANOVA table for Mean SR at 95% confidence level. The data for each factor and its interaction were F tested, Where F is the Fister Value. The greater its effect on the performance characteristics. The corresponding values in ANOVA table are shown in table 4.14. This table shows that pulse off time, current, pulse on time affects the SR significantly. Pulse on has the highest contribution to SR.

4.3.4 Confirmation test for SR

From mean of each level of every factor, response table is constructed for SR table no 4.12 and 4.13. From Table it is concluded that for SR, the optimum conditions are A1, B1 and C2 i.e. current 6 A, Pulse on 50 μs and pulse off 6 μs. From figure 4.5 and 4.6, shows the main effect plot for S/N ratio and Means and obtaining optimal values.

Graph show that larger the value of parameters better the SR. Here A is current; min at A1, pulse on; max at B1, pulse off C2. The ANOVA table is shown in table 4.14

Table 4.12 Response table for S/N ratio

Response table for S/N ratio			
Level	Current (A)	Pulse On (B)	Pulse off (C)
1	-13.72	-12.21	-14.66
2	-14.18	-15.24	-14.31
3	-15.63	-16.08	-14.57
Delta	1.92	3.87	0.35
Rank	2	1	3

From graph 4.5 and 4.6, main effects plot for SR conclude that optimum condition for SR is Current (6 A), Pulse on (50 μs), Pulse off (6 μs). The main effects plot shows the values of parameters and also showing optimal values of parameters. Plots are explained in the respective heading. From plots we can conclude the optimum value of the parameters to make the process robust.

Table 4.13 Response Table for Means

Response Table for Means			
Level	Current (A)	Pulse On (B)	Pulse off (C)
1	4.961	4.117	5.624
2	5.241	5.814	5.325
3	6.11	6.38	5.363
Delta	1.149	2.263	0.3
Rank	2	1	3

Table 4.14 – ANOVA Table for S/N Ratio

Analysis of Variance for SN ratio					
Source	DF	Seq. SS	Adj. MS	F	P
Current	2	6.0115	3.0058	4.14	0.194
Pulse on	2	24.9158	12.4579	17.18	0.055
Pulse off	2	0.1989	0.0995	0.14	0.879
Residual Error	2	1.4506	0.7253		
Total	8	32.5769			

Table 4.15 – ANOVA Table for Means

Analysis of Variance for Means					
Source	DF	Seq. SS	Adj. MS	F	P
Current	2	2.1535	1.07675	6.21	0.139
Pulse on	2	8.3207	4.16035	24.01	0.04
Pulse off	2	0.1596	0.07979	0.46	0.685
Residual Error	2	0.3466	0.17329		
Total	8	10.9804			

For SR, ANOVA is smaller the better. From table 4.12, 4.13, it is concluded that pulse on is the most significant factor for SR, followed by current and then pulse off. So pulse off is the least significant factor.

Main effects plot for SR conclude that optimum condition for SR is Current (6 A), Pulse on (50 μs), Pulse off (8 μs). The main effects plot shows the values of parameters and also showing optimal values of parameters.

4.3.5 Main effects plots for SR of AISI D2 steel

We can see that the graphs are showing minimum as well as maximum values of parameters. For SR, we chosen ANOVA as smaller is better so here we are interested in minimum values. There are two plots; one is for SN ratio and other is for means. For SN ratio we are always interested in maximum values because it tells us which parameter at which value has the maximum effects on the process. Whereas for means plot, we take mean of the values, here smaller is better so smaller values are considered to optimize the process. As we can see in the graphs below, we see the values which we can use to achieve minimum surface roughness. For SN graph maximum values are taken and for Mean of means graph minimum values are taken. So achieve high surface finish or low surface roughness, EDM can be done using the parameter values Current (6 A), Pulse on (50 µs), Pulse off (8 µs).

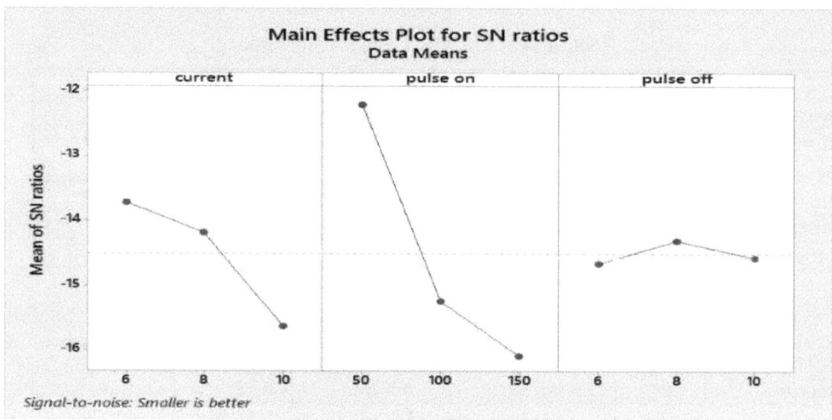

Figure 4.5-Main Effects Plot for SN ratio

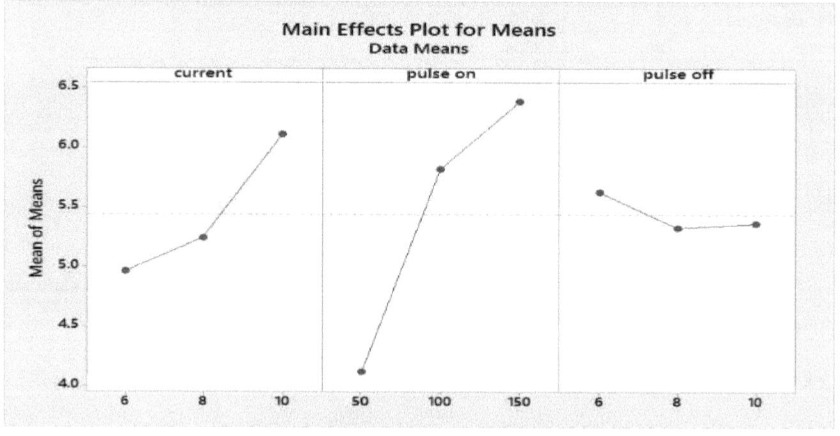

Figure 4.6-Main Effects plot for Means

4.4 Analysis and result of MRR of SS 316

Total 9 experiments were conducted for the L_9 orthogonal array for both the materials. The results of each experiment were repeated three times, which means for every experiment there are three values of material removal rate. So the analysis is based on it. But to calculate results of MRR, S/N ratio is to be calculated first.

4.4.1 Analysis and result of MRR for machined surface

The effect of input parameters i.e. current, Pulse on time, pulse off time was evaluated using ANOVA. Signal to noise ratio is calculated and the response data specially used for design of experiments application is analyzed.

(S/N) HB = -10 log (MSDHB)

Where MSDHB=1/r$\sum_{i}^{r} = (\frac{1}{y_i^2})$

r = the number of tests in a trial

y_i = observed value of response characteristics

For material removal rate (MRR) the S/N ratio is larger is better.

Where, MSDHB = Mean Square deviation for higher the better response.

4.4.2 Observation Table for MRR

While conducting all the 9 experiments with different values of input parameters, observations were made. Weight lost per gram from the workpiece in each experiment and time taken for each experiment was also observed. The depth of cut was fixed to 0.5 mm and time taken to achieve this cut was observed. For some experiments time was higher and for some experiments time is low. So we calculated an average material removal rate for 10 minutes. And after each experiment the weight of workpiece is measured and loss in weight is observed as removal of material. After completion of all experiments the observation table 4.16 is made by filling all the values.

$$\text{MRR} = \frac{\text{Final weight} - \text{Initial weight}}{\text{Time of machining}}$$

Table 4.16 Observation Table for Material removal Rate for SS316

Serial No	Current Ip, A	Pulse on Ton, μs	Pulse off Toff, μs	Weight of workpiece (Grams)			SNRA1	MEAN1
				MRR1	MRR2	MRR3		
1	6	50	6	0.0469	0.0458	0.046	-26.70227932	0.046233333
2	6	100	8	0.207	0.1998	0.2109	-13.73343988	0.2059
3	6	150	10	0.4125	0.4215	0.4225	-7.56069199	0.418833333
4	8	50	8	0.1257	0.1156	0.1205	-18.38828096	0.1206
5	8	100	10	0.4076	0.411	0.41	-7.75437334	0.409533333
6	8	150	6	0.1439	0.1552	0.1502	-16.50419636	0.149766667
7	10	50	10	0.725	0.7325	0.7199	-2.784321661	0.7258
8	10	100	6	0.2478	0.2564	0.2509	-11.98492108	0.2517
9	10	150	8	0.4928	0.4876	0.49	-6.193959985	0.490133333

4.4.3 Analysis of Variance for MRR

The results were analyzed using ANOVA for identifying the significant factors affecting performance. The variation data for each factor and their interactions were tested to find significance of each calculated by formula. Main effect plot is shown in fig. 4.7, shows the variation of MRR with input parameters. The ANOVA table for Mean MRR at 95% confidence level. The data for each factor and its interaction were F tested, Where F is the Fister Value. The greater its effect on the performance characteristics. The corresponding values in ANOVA table are shown in table 4.19. This table shows that pulse off time, current, pulse on time affects the MRR significantly. Pulse off has the highest contribution to MRR.

4.4.4 Confirmation test for MRR

From mean of each level of every factor, response table is constructed for MRR table no 4.17 and 4.18. From Table it is concluded that for MRR, the optimum conditions are A3, B2 and C3 i.e. current 10 A, Pulse on 100 μs and pulse off 10 μs. From Graph 4.7 and 4.8, shows the main effect plot for S/N ratio and Means and obtaining optimal values.

Graph show that larger the value of parameters better the MRR. Here A is current; max at A3, pulse on; max at B2, pulse off C3. The ANOVA table is shown in table 4.19. From graph 4.7 and 4.8, main effects plot for MRR conclude that optimum condition for MRR is Current (10 A), Pulse on (150 μs), Pulse off (10 μs). The main effects plot shows the values of parameters and also showing optimal values of parameters.

Table 4.17 Response Table for S/N ratio

Response table for S/N ratio			
Level	Current (A)	Pulse On (B)	Pulse off (C)
1	-15.999	-15.958	-18.397
2	-14.216	-11.158	-12.772
3	-6.988	-10.086	-6.033
Delta	9.011	5.872	12.364
Rank	2	3	1

Table 4.18 Response table for Means

Response Table for Means			
Level	Current (A)	Pulse On (B)	Pulse off (C)
1	0.2237	0.2975	0.1492
2	0.2266	0.289	0.2722
3	0.4892	0.3529	0.5181
Delta	0.2656	639	0.3688
Rank	2	3	1

Table 4.19 – ANOVA Table for S/N Ratio

Analysis of Variance for S/N Ratio					
Source	DF	Seq. SS	Adj. MS	F	P
Current	2	136.621	68.311	13.71	0.068
Pulse on	2	58.675	29.338	5.89	0.145
Pulse off	2	229.923	114.961	23.07	0.042
Residual Error	2	9.967	4.984		
Total	8	435.186			

Table 4.20 – ANOVA Table for Means

Analysis of Variance for Means					
Source	DF	Seq. SS	Adj. MS	F	P
Current	2	0.139476	0.069738	9.91	0.092
Pulse on	2	0.007217	0.003608	0.51	0.661
Pulse off	2	0.211593	0.105796	15.04	0.062
Residual Error	2	0.014069	0.007035		
Total	8	0.372355			

4.4.5 Main effects plots for MRR of SS 316

We can see that the graphs are showing minimum as well as maximum values of parameters. For MRR, we chosen ANOVA as larger is better so here we are interested in maximum values. There are two plots; one is for SN ratio and other is for means. For SN ratio we are always interested in maximum values because it tells us which parameter at which value has the maximum effects on the process. Whereas for means plot it depends on ANOVA, here larger is better so larger are considered to optimize the process.

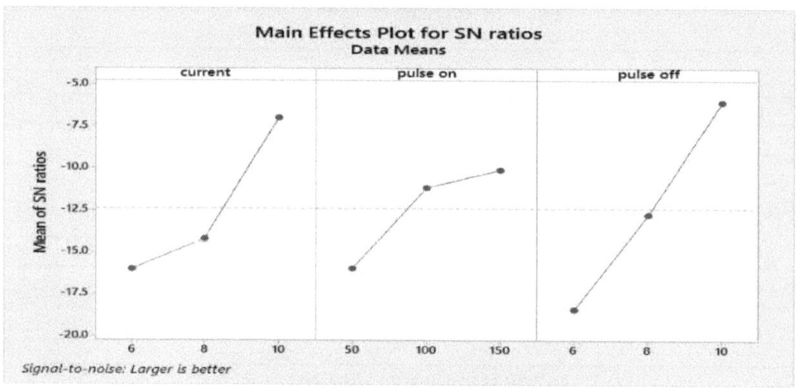

Figure 4.7-Main Effects plot for SN ratio

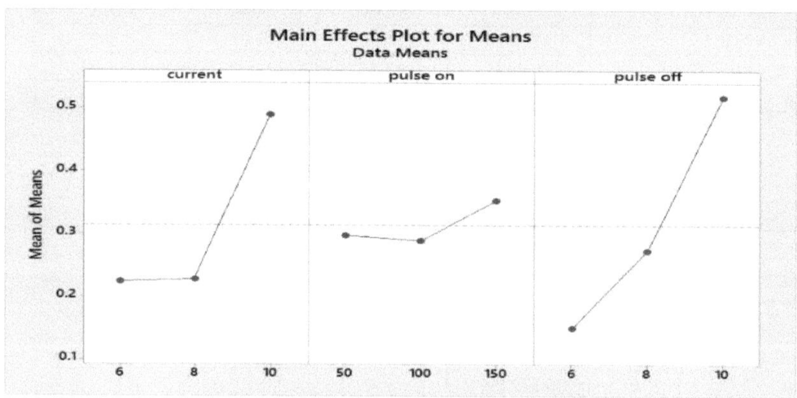

Figure 4.8-Main effects plot for Means

4.5 Analysis and Results of TWR of SS 316

Total 9 experiments were conducted for the L_9 orthogonal array for both the materials. The results of each experiment were repeated three times, which means for every experiment there are three values of material removal rate. So the analysis is based on it. But to calculate results of TWR, S/N ratio is to be calculated first.

4.5.1 Results and analysis of TWR

The effect of input parameters i.e. current, Pulse on time, pulse off time was evaluated using ANOVA. Signal to noise ratio is calculated and the response data specially used for design of experiments application is analyzed.

(S/N) LB = -10 log (MSDLB)

Where, MSDLB=$1/r\sum_{i=1}^{r} = (y_i^2)$,

r = the number of tests in a trial

yi = observed value of response characteristics

For Tool wear rate (TWR) the S/N ratio is smaller is better.

Where MSDLB = Mean Square deviation for smaller the better response.

4.5.2 Observation Table for TWR

While conducting all the 9 experiments with different values of input parameters, observations were made. Weight lost per gram from the electrode in each experiment and time taken for each experiment was also observed. The depth of cut on workpiece was fixed to 0.5 mm and time taken to achieve this cut was observed.

For some experiments time was higher and for some experiments time is low. So we calculated an average tool wear rate for 10 minutes. And after each experiment the weight of tool electrode is measured and loss in weight is observed as removal of material. After completion of all experiments the observation table 4.21 is made by filling all the values.

The observation table is shown below, showing the values of signal to noise ratio and means. These values are calculated by software. To calculate the signal to noise value and mean value, we have to first input values of TWR into the software and based on this values, the software will calculate mean values

Table 4.21 Observation Table for Tool Wear Rate of SS316

Serial No	Current Ip, A	Pulse on Ton, μs	Pulse off Toff, μs	Weight of electrode (Grams) TWR1	TWR2	TWR3	SNRA1	MEAN1
1	6	50	6	0.0168	0.0159	0.0163	35.73629731	0.016333333
2	6	100	8	0.0222	0.0278	0.0243	32.08503506	0.024766667
3	6	150	10	0.0187	0.0169	0.0174	35.04890637	0.017666667
4	8	50	8	0.02	0.022	0.0236	33.18469329	0.021866667
5	8	100	10	0.0615	0.0572	0.0589	24.54969303	0.0592
6	8	150	6	0.0146	0.0151	0.0162	36.29789252	0.0153
7	10	50	10	0.125	0.128	0.124	18.0148037	0.125666667
8	10	100	6	0.026	0.0225	0.024	32.32040629	0.024166667
9	10	150	8	0.0571	0.053	0.058	25.02452544	0.056033333

4.5.3 Analysis of Variance

The results were analyzed using ANOVA for identifying the significant factors affecting performance. The variation data for each factor and their interactions were F tested to find significance of each calculated by formula. Main effect plot is shown in fig. 4.9, shows the variation of TWR with input parameters. The ANOVA table for Mean TWR at 95% confidence level. The data for each factor and its interaction were F tested, Where F is the Fister Value. The greater its effect on the performance characteristics. The corresponding values in ANOVA table are shown in table 4.24. This table shows that pulse off time, current, pulse on time affects the TWR significantly. Current has the highest contribution to TWR.

4.5.4 Confirmation test for TWR

From mean of each level of every factor, response table is constructed for TWR table no 4.22 and 4.23. From Table it is concluded that for TWR, the optimum conditions are A1, B3 and C1 i.e. current 6 A, Pulse on 150 μs and pulse off 6 μs.

Table 4.22 Response table for S/N ratio

Response table for S/N ratio			
Level	Current (A)	Pulse On (B)	Pulse off (C)
1	34.29	28,98	34.78
2	31.34	29.65	30.1
3	25.12	32.12	25.87
Delta	9.17	3.15	8.91
Rank	1	3	2

From Graph 4.9 and 4.10, shows the main effect plot for S/N ratio and Means and obtaining optimal values. Graph show that larger the value of parameters better the TWR. Here A is current; max at A1, pulse on; max at B3, pulse off C1. The ANOVA table is shown in table 4.25.

Main effects plot for TWR conclude that optimum condition for MRR is Current (6 A), Pulse on (150 μs), Pulse off (6 μs). The main effects plot shows the values of parameters and also showing optimal values of parameters.

Table 4.23 Response table for Means

Response Table for Means			
Level	Current (A)	Pulse On (B)	Pulse off (C)
1	0.01959	0.05462	0.0186
2	0.03212	0.03603	0.03422
3	0.06862	0.022967	0.06751
Delta	0.04903	0.02496	0.04891
Rank	1	3	2

Table 4.24 – ANOVA Table for S/N Ratio

Analysis of Variance for S/N ratio					
Source	DF	Seq. SS	Adj. MS	F	P
Current	2	131.51	65.756	2.73	0.268
Pulse on	2	16.46	8.228	0.34	0.746
Pulse off	2	119.29	59.644	2.47	0.288
Residual Error	2	48.22	24.108		
Total	8	315.47			

Table 4.25 – ANOVA Table for Means

Analysis of Variance for Means					
Source	DF	Seq. SS	Adj. MS	F	P
Current	2	0.003894	0.001947	2.17	0.316
Pulse on	2	0.001009	0.000504	0.56	0.641
Pulse off	2	0.003745	0.001872	2.08	324
Residual Error	2	0.001798	0.000899		
Total	8	0.010445			

For TWR, ANOVA is smaller the better. From table 4.24, 4.25, it is concluded that current is the most significant factor for TWR, followed by pulse off and then pulse on. So pulse on is the least significant factor.

From graph 4.9 and 4.10, main effects plot for TWR conclude that optimum condition for TWR is Current (6 A), Pulse on (150 μs), Pulse off (6 μs). The main effects plot shows the values of parameters and also showing optimal values of parameters.

4.5.5 Main effects plots

We can see that the graphs are showing minimum as well as maximum values of parameters. For TWR, we chosen ANOVA as smaller is better so here we are interested in minimum values. There are two plots; one is for SN ratio and other is for means.

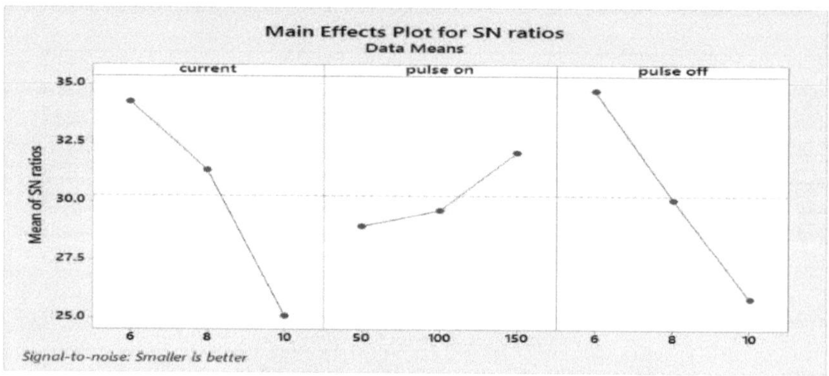

Figure 4.9-Main Effects for SN ratios

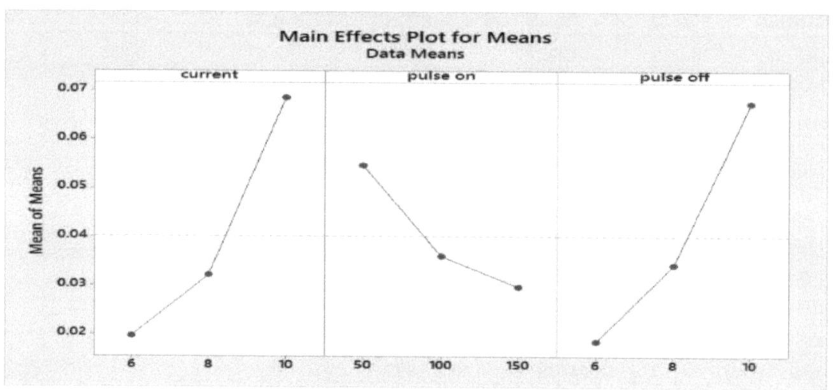

Figure 4.10-Main Effects for Means

4.6 Results and Analysis of Surface Roughness of SS316

Total 9 experiments were conducted for the L_9 orthogonal array for both the materials. The results of each experiment were repeated three times, which means for every experiment there are three values of surface roughness. So the analysis is based on this experimentation table. To calculate results of SR, S/N ratio is to be calculated first.

4.6.1 Results and analysis of SR

The effect of input parameters i.e. current, Pulse on time, pulse off time was evaluated using ANOVA. Signal to noise ratio is calculated and the response data specially used for design of experiments application is analyzed.

(S/N) LB = -10 log (MSDLB)

Where, MSDLB=$1/r\sum_{i=1}^{r} = (y_i^2)$,

r = the number of tests in a trial

y_i = observed value of response characteristics

For Tool wear rate (TWR) the S/N ratio is smaller is better.

Where MSDLB = Mean Square deviation for smaller the better response.

4.6.2 Observation Table for SR

While conducting all the 9 experiments with different values of input parameters, observations were made. Weight lost per gram from the workpiece in each experiment and time taken for each experiment was also observed. The depth of cut was fixed to 0.5 mm and time taken to achieve this cut was observed. For some experiments time was higher and for some experiments time is low. So we calculated an average material removal rate for 10 minutes. And after completion of experimental work, Surface roughness is measured of these cuts. After completion of all experiments the observation table 4.26 is made by filling all the values.

Table 4.26 Observation Table for Surface Roughness of SS316

Serial No	Current Ip, A	Pulse on Ton, µs	Pulse off Toff, µs	Weight of electrode (Grams)			SNRA1	MEAN1
				SR1	SR2	SR3		
1	6	50	6	4.534	4.243	4.432	-12.8782284	4.403
2	6	100	8	5.133	5.698	5.443	-14.69533087	5.424666667
3	6	150	10	6.149	5.745	6.253	-15.63936638	6.049
4	8	50	8	4.378	4.135	4.051	-12.44489143	4.188
5	8	100	10	7.297	7.546	7.069	-17.27430327	7.304
6	8	150	6	6.488	6.143	6.376	-16.03808011	6.335666667
7	10	50	10	6.045	6.316	6.431	-15.93993221	6.264
8	10	100	6	6.618	6.754	6.552	-16.44580193	6.641333333
9	10	150	8	7.661	7.211	7.521	-17.46257551	7.464333333

4.6.3 Analysis of Variance for SR

The results were analyzed using ANOVA for identifying the significant factors affecting performance. The variation data for each factor and their interactions were tested to find significance of each calculated by formula. Main effect plot is shown in fig. 4.11, shows the variation of SR with input parameters. The ANOVA table for Mean SR at 95% confidence level. The data for each factor and its interaction were F tested, Where F is the Fister Value. The greater its effect on the performance characteristics. The corresponding values in ANOVA table are shown in table 4.29. This table shows that pulse off time, current, pulse on time affects the SR significantly. Pulse on has the highest contribution to SR.

4.6.4 Confirmation test for SR

From mean of each level of every factor, response table is constructed for SR table no 4.27 and 4.28. From Table it is concluded that for SR, the optimum conditions are A1, B3 and C2 i.e. current 6 A, Pulse on 150 µs and pulse off 8 µs. From Graph 4.11 and 4.12, shows the main effect plot for S/N ratio and Means and obtaining optimal values.

Graph show that larger the value of parameters better the SR. Here A is current; max at A3, pulse on; max at B2, pulse off C3. The ANOVA table is shown in table 4.30

Table 4.27 Response Table for S/N ratio

Response table for S/N ratio			
Level	Current (A)	Pulse On (B)	Pulse off (C)
1	-14.4	-13.75	-15.12
2	-15.25	-16.14	-14.87
3	-16.62	-16.38	-16.28
Delta	2.21	2.63	1.42
Rank	2	1	3

From graph 4.11 and 4.12, main effects plot for SR conclude that optimum condition for SR is Current (10 A), Pulse on (100 µs), Pulse off (10 µs). The main effects plot shows the values of parameters and also showing optimal values of parameters. Plots are explained in the respective heading. From plots we can conclude the optimum value of the parameters to make the process robust.

Table 4.28 Response Table for Means

Response Table for Means			
Level	Current (A)	Pulse On (B)	Pulse off (C)
1	5.292	4.952	5.793
2	5.943	6.457	5.692
3	6.79	6.612	6.539
Delta	1.498	1.665	0.847
Rank	2	1	3

Table 4.29 – ANOVA Table for S/N Ratio

Analysis of Variance for S/N ratio					
Source	DF	Seq. SS	Adj. MS	F	P
Current	2	7.471	3.7355	4.48	0.182
Pulse on	2	12.636	6.3182	7.58	0.117
Pulse off	2	3.426	1.7131	2.06	0.327
Residual Error	2	1.666	0.8332		
Total	8	25.2			

Table 4.30 – ANOVA Table for Means

Analysis of Variance for Means					
Source	DF	Seq. SS	Adj. MS	F	P
Current	2	3.3839	1.692	3.86	0.206
Pulse on	2	5.0616	2.5308	5.78	0.148
Pulse off	2	1.2831	0.6415	1.46	0.406
Residual Error	2	0.8763	0.4382		
Total	8	25.2			

For SR, ANOVA is smaller the better. From table 4.29, 4.30, it is concluded that pulse on is the most significant factor for SR, followed by current and then pulse off. So pulse off is the least significant factor.

Main effects plot for SR conclude that optimum condition for TWR is Current (6 A), Pulse on (150 μs), Pulse off (8 μs). The main effects plot shows the values of parameters and also showing optimal values of parameters.

4.6.5 Main effects plots for SR of SS 316

From main effects plots 4.11 and 4.12, we can see that the graphs are showing minimum as well as maximum values of parameters. For SR, we chosen ANOVA as smaller is better so here we are interested in minimum values. There are two plots; one is for SN ratio and other is for means. For SN ratio we are always interested in maximum values because it tells us which parameter, at which value has the maximum effects on the process.

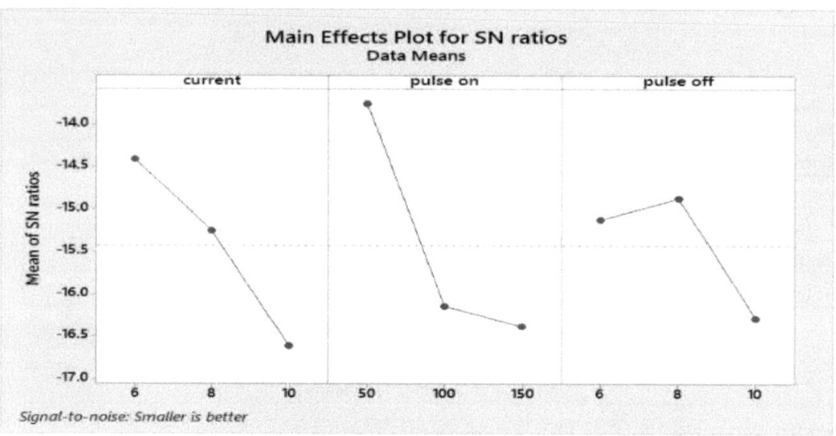

Figure 4.11-Main Effects for SN ratio

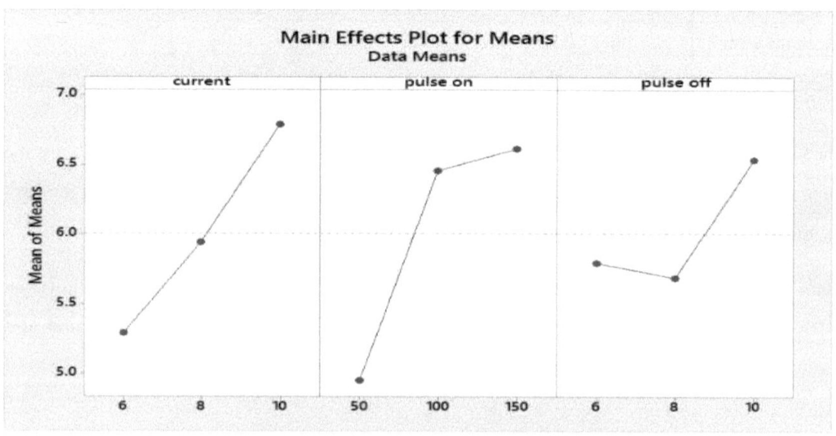

Figure 4.12-Main effects plot for Means

Whereas for means plot, wc takc mean of the values, here smaller is better so smaller values are considered to optimize the process. As we can see in the graphs below, we see the values which we can use to achieve minimum surface roughness. For SN graph maximum values are taken and for Mean of means graph minimum values are taken. So achieve high surface finish or low surface roughness, EDM can be done using the parameter values Current (6 A), Pulse on (50 μs), Pulse off (8 μs).

Conclusion

The objective of the present study is to investigate the effect of input parameters namely current, pulse on time and pulse off time on output parameters namely MRR, TWR and SR using aluminium electrode. According to V.D. Patel (21), at the same current, aluminium electrode yielded the highest MRR. This study was done on Mild Steel. But in present study higher grades of steel are considered. L9 orthogonal array based on Taguchi method and ANOVA was performed for analyzing the result.

The following conclusions has been made:

1. For MRR, the most influencing factor is pulse off time, followed by current and then pulse on time for both high carbon and high chromium steel and SS 316. The Optimum conditions for MRR is Current (10 A), Pulse on (100 μs), Pulse off (10 μs) for high carbon high chromium steel. The optimum condition for SS 316 for MRR is Current (10 A), Pulse on (150 μs), Pulse off (10 μs).
2. For TWR, the most influencing factor is pulse off time, followed by current and then pulse on time. The optimum condition for TWR is Current (6A), Pulse on (150 μs), Pulse off (6 μs) for high carbon high chromium steel. For SS316, for TWR, Current is the most influencing factor followed by pulse off and then pulse on. The optimum condition is Current (6A), Pulse on (150 μs), Pulse off (6 μs).
3. For SR, pulse on time is the most influencing factor followed by current and pulse off time. The optimum condition for SR is current (6A), Pulse on (50 μs), Pulse off (8 μs) for both high carbon high chromium steel and SS 316.

There was very little to no research work has been reported on the use of aluminium as tool electrode. Depending on the tool electrode selection criteria given in this study, aluminium can be used as tool electrode and experimental investigation in the current work proves that aluminium can be successfully used as a tool electrode in EDM.

Scope for Future work

Some parameters like Voltage, dielectric, polarity are kept constant. These can be varied and their effect can be studied. Now a days, environmental safety is a big issue, so instead of using hydrocarbon oil, Distilled water can be used as dielectric and same experimentation can be repeated to see the EDM performance in water. Dry EDM can also be done. PMEDM can also be performed, for e.g. mixing any of the powder of graphite, chromium, aluminium, tungsten in dielectric and further optimizing the process. Powder mixed EDM is very popular now a days because it improves the performance of EDM significantly. Generally, a conductive powder is used to mix in dielectric but recently researchers started to mix non-conductive powder and studying its effects.

References

1. Abbas, N.M., Solomon, D.G. and Bahari, M.F. (2007),"A review on current research trends in electrical discharge machining (EDM)", International Journal of Machine Tools and Manufacture, Vol 47, Issue 1, pp. 1214-1228

2. Amin, N., Lajis, M.A. and Radzi, M. (2009),"The Implementation of Taguchi method on EDM process of Tungsten carbide", European Journal of Scientific Research, Vol 26,Issue 1, pp. 609-617

3. Choudhary, S., Kant, K. and Saini, P. (2013),"Analysis of MRR and SR with different electrode for SS316 on die sinking EDM using Taguchi technique", Global Journal of Researches in Engineering Mechanical and Mechanics Engineering, Vol 13, Issue 3, pp. 112-120

4. Cogun, C. and Karacay, T. (2006)," An experimental investigation on the effect of powder mixed dielectric on machining performance in electric discharge machining performance in electric discharge machining", Journal of Engineering Manufacture, Vol 220, Part B, pp 1035-1049

5. Dhanabalan, S., Sivakumar, K and Narayanan, C.S. (2015),"Experimental investigation on electrical discharge machining of titanium alloy using copper, brass and aluminium electrodes", Journal of Engineering Science and Technology, Vol 10, Issue 1, pp 72-80

6. Dixit, A.C., Kumar, A., Singh, R.K. and Bajpai, R. (2015),"An experimental study of material removal rate and electrode wear rate of high carbon high chromium steel (AISI D3) in EDM process using copper tool electrode", International Journal of Innovative Research in Advanced Engineering, Vol 3, Issue 1, pp 257-262

7. Gopalakannan, S. and Senthivelan, T. (2012),"Effect of electrode materials on electrical discharge machining of 316L and 17-4 PH stainless steel", Journal of Minerals and Materials Characteristics and Engineering, Vol 11,Issue 3, pp. 685-690

8. Gudur, S and Potdar, V.V. (2015),"Effect of silicon carbide powder mixed EDM on machining characteristics of SS316L material", International Journal of Innovation Research in Science, Engineering and Technology, Vol 4. Issue 9, pp. 8131-8141

9. Haron, C.H., Ghani, J.A., Burhamidin, Y. and swee, C.Y. (2008),"Copper and graphite electrodes performance in electrical discharge machining of XW42 tool steel", Journal of Material Processing Technology, Vol 201, Issue 2, pp. 570-573

10. Hindus. S. J., Kumar, R., P.R., oppiliyappan, B and kuppan, P. (2013),"Experimental investigations on electrical discharge machining of SS316L", International Journal on Mechanical Engineering and Robotics, Vol 1, Issue 2, pp. 481-488

11. Ho, K.H. and Newman, S.T. (2003),"State of the art electrical discharge machining (EDM)", International Journal of Machine Tools and Manufacture, Vol 43, Issue5, pp. 1287-1300

12. Hu, F.Q., Cao, F.Y., Song, B.Y., Hou, P.J., Zhang, Y., Chen, K. and Wei J.Q. (2013),"Surface properties of SiCp/Al composite by powder mixed EDM", Procedia CIRP 6 101-106

13. Jamadar, M.M. and Kavade, M.V. (2014),"Effect of aluminium powder mixed EDM on machining characteristics of die steel (AISI D3)", Proceedings of 10th IRF international conference, pp 120-123

14. Jegan, T.M.C., M., Dev A., M. and Ravindran, D. (2012),"Determination of electro discharge machining parameters in AISI 202 stainless steel using grey relational analysis", International Conference on Modelling Optimization and Computing, Volume 38, pp 4005-4012

15. Jianxin, D. and Taichiu, L. (2002), "Effect of ultrasonic surface finishing on the strength and thermal shock behavior of the EDM ceramic composite", International Journal of Machine Tools & Manufacture, Vol.42, Issue 4, pp. 245-250.

16. Kansal, H.K., Singh, S. and Kumar, P. (2006),"Performance parameters optimization (multi-characteristics) of powder mixed and utility concept," Indian Journal of Engineering and Material Sciences, Vol 13, Issue 3, pp 209-216

17. Khanra, A.K., Sarjar, B.R., Bhattacharya, B., Pathak, L.C. and Godkhindi, M.M. (2007),"Performance of ZrB_2-Cu composite as an EDM electrode," Journal of Materials Processing Technology, Vol 183, Issue 3, pp. 122-126

18. Kristian, A.L. (2004)," Performance of two graphite qualities in EDM of Steel slots in a jet engine turbine vane", Journal of Materials Processing Technology, Vol 149, Issue 3pp. 152-158

19. Lauwers, B., Krath, J.P., Liu, W., Earaets, W., Schacht, B. and Bleys, P. (2004),"Investigation of materials removal mechanisms in EDM of composite ceramic material", Journal of Materials Processing Technology, Vol 149, pp. 347-352

20. Laxman, J. and Gururaj, K. (2014),"Optimization of EDM process parameters on titanium super alloys based on the grey relational analysis", International Journal of Engineering Research, Volume 3, Issue 5, pp 344-348

21. Lee, S.H. and Li, X.P. (2001),"Study of the machining parameters on the machining characteristics in electrical discharge machining of Tungsten Carbide", Journal of Materials Processing Technology, Vol 115, pp. 344-358

22. Mohd. J.M., Khalid S., Singh B. and Malhotra N. (2012), "Modeling and analysis of machining parameters for surface roughness in powder mixed EDM using RSM approach"; International Journal of Engineering, Science and Technology Vol. 4, No. 3, pp. 45- 52.

23. Mumiu, J.M., Ikua, B.W., Nyaanga, D.M. and Gieharu, S.N. (2013),"applicability of diatomic powder-mixed dielectric fluid in electrical discharge machining processes", International Journal of Pure and Applied Sciences and Technology, pp 19-36

24. Murikan, R.T., Jakkamputi, L.P. and Kuppan, P. (2013),"Experimental investigation of dry electrical discharge machining on SS 316L", International Journal of Latest Trends in Engineering and Technology, Vol 2, Issue 3

25. Khundrakpam N., S., Kumar, S., Singh, A., Brar G., S. (2014), "Study and Analysis of Zinc PMEDM Process Parameters on MRR"; International Journal of Emerging Technology and Advanced Engineering, Volume 4, Issue 3.

26. Ojha, K., Garg, R.K. and Singh, K.K. (2011),"Parametric optimization of PMEDM process using chromium powder mixed Dielectric and triangular shape electrodes", Journal of Minerals and Materials Characteristics and Engineering, Vol 10, pp 1087-1102

27. Patel, V.D. Mr., Patel, C.P. Prof, and Patel, V.J. ,"Analysis of different tool materials on MRR and surface roughness of Mild steel in EDM", International Journal of Engineering Research and Application, Vol 1, Issue 3, pp 394-397

28. Rahi, D.K., and Vishwakarma, M. (2014),"Performance measurement of EDM parameters on high carbon high chromium steel", International Journal of Engineering Research and Development, Vol 10, Issue 7, pp 15-21

29. Rajendra, M. and Rao, K.M. (2014),"Experimental evaluation of performance of electrical discharge machining of D3 steel with Al_2O_3 abrasive mixed dielectric material by using design of experiments", International Journal of Research in Engineering and Technology, Vol 3, issue 1, pp. 599-606

30. Raju, P., Sarcar, M.M.M. and Satyanarayana, B.(2014),"Optimization of wire electric discharge machining parameters for surface roughness on SS316L stainless steel using full factorial experimental design", International Conference on Advances in Manufacturing and Materials Engineering, AMME 2014, Procedia material sciences, pp. 1670-1676

31. Sameh S. H. (2009),"Study of the parameters in electrical discharge machining through response surface methodology approach", Journal of Applied Mathematical Modelling, Vol 10, Issue 2, pp 4397-4407

32. Shivam, Rakesh , (2014), "Parametric Study of Powder Mixed EDM and Optimization of MRR & Surface Roughness", International Journal of Scientific Engineering and Technology, Vol 3, Issue 1, pp. 427-438

33. Shriram Y. Kaldhone and M.V. Kavade,(2014), " An Experimental Study on the Effect of Powder Mixed Dielectric On Machining Performance In Electric Discharge Machining of Tungsten Carbide"; Proceedings of 12th IRF International Conference.

34. Singh, S and Bhardwaj, A. (2011),"Review to EDM by using water and powder mixed dielectric fluid", Journal of Minerals and Materials Characterization and Engineering, Vol 10, Issue 2, pp. 199-230

35. Singh, H. and Singh, A. (2012),"Effect of pulse on/pulse off on machining of AISI D3 die steel using copper and brass electrode in EDM", International Journal of engineering and science, Vol 1. Issue 9, pp 19-22

36. Suresh, P., R., Venketesan, T., Sekar and Sathiyamoorthy, V. (2014),"Study of microEDM parameters of stainless steel 316L: Material removal rate optimization using genetic algorithm", International Journal of Engineering and Technology, Vol 6, Issue 2, pp 1065-1071

37. Syed, K.H. and Palaniyandi, K. (2012),"Performance of electrical discharge machining using aluminium powder suspended distilled water", Turkish Journal of Engineering and Environmental Science, Vol 36, Issue 1, pp 195-207